The Greatest Physicists of All Time

A Visual and Poetic Guide

Preetinder Rahil

The author may have checked for errors twice or even thrice,
but doing your due diligence would still be considered wise.

The Greatest Physicists of All Time

Galileo Galilei

Isaac Newton

Michael Faraday

James Maxwell

Ludwig Boltzmann

Max Planck

Marie Curie

Albert Einstein

Ernest Rutherford

Niels Bohr

Louis De Broglie

Wolfgang Pauli

Erwin Schrödinger

Max Born

Werner Heisenberg

Paul Dirac

Subrahmanyan Chandrasekhar

Enrico Fermi

Richard Feynman

Murray Gell-Mann

Steven Weinberg

Galileo Galilei

Galileo was a man of the Renaissance.
A man of mathematical rigor who left nothing
to chance.

In the scientific revolution's start,
Galileo played an important part.

1

From the leaning Tower of Pisa
things may not have been dropped.
But his work on inertia and mechanics
could not be stopped.

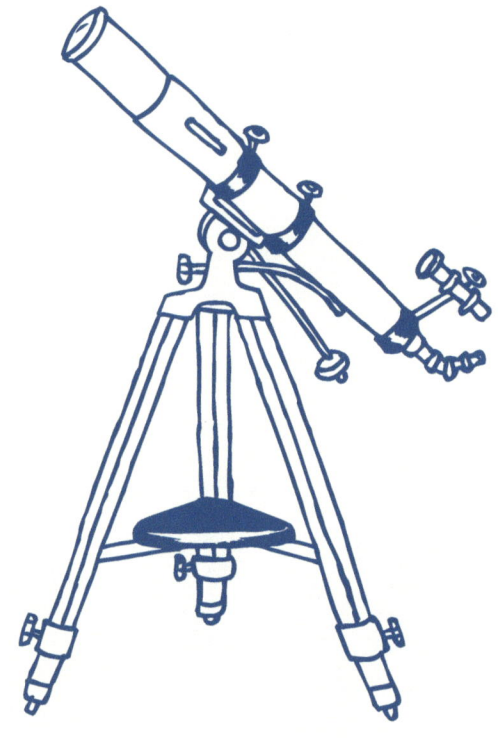

With his telescope, he could see quite far.
The sunspots, moon craters, and glows of a distant star.

He advocated that the earth revolves around the sun.
Aristotle's theories were obsolete and done.

3

It was a heresy that the orthodoxy couldn't digest.
He was sentenced and put under house arrest.

He may have fought a lonely fight.
But time has eventually proved him right.

Isaac Newton

If we hadn't known Newton's name,
history would not have been the same.

His accomplishments are all well-known.
With time, his stature has only grown.

When an apple fell on the ground.
Newton thought a reason must be found.

Newton said a force must be there.
And must be present everywhere.

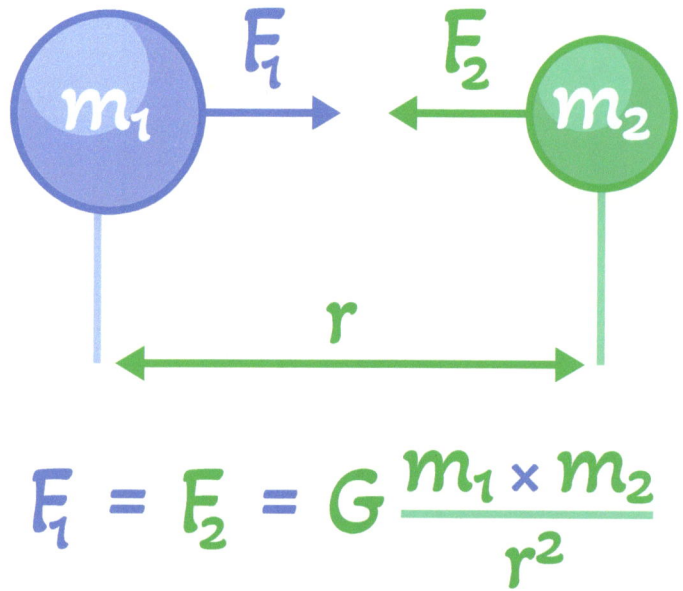

$$F_1 = F_2 = G\,\frac{m_1 \times m_2}{r^2}$$

Planets movements and the apple's fall
were united by gravity once and for all.

After much contemplation,
Newton came up with his law of gravitation.

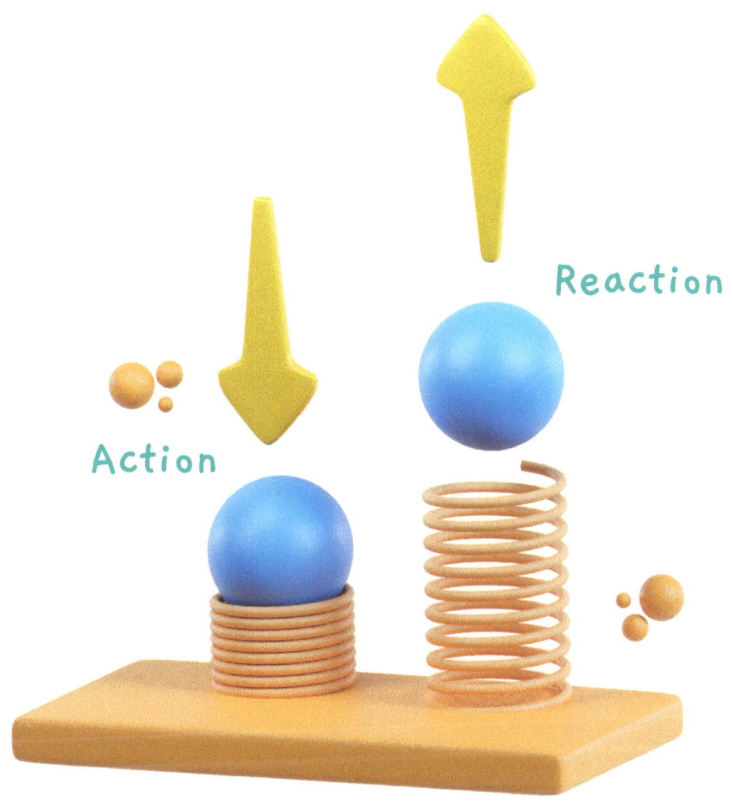

Why do things move and why are they at rest?
Newton's laws of motion have withstood every test.

His book, the Principia Mathematica is much prized,
where his brilliant theorems are summarized.

8

Newton showed light was not just white.
The colors of the visible spectrum couldn't escape his sight.

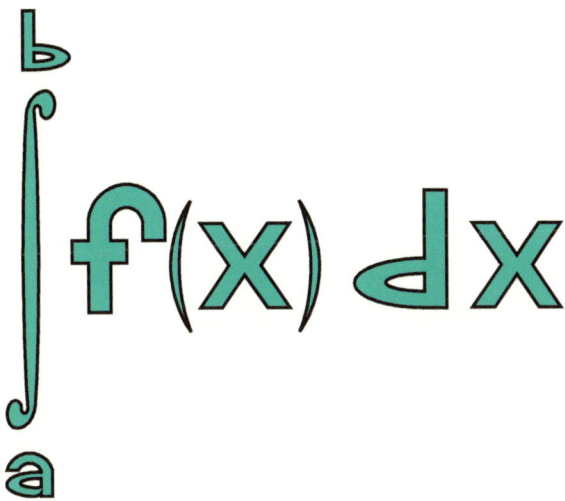

$$\int_a^b f(x)\, dx$$

He also studied the mathematics of continuous change.
Developing calculus was within his intellectual range.

Working in alchemy showed Newton's flaws.
A man of contradiction, that's what he was.

Michael Faraday

He discovered electromagnetic induction
where changing electric fields lead to
electricity production.

It would indeed be fair
to call him the experimentalist extraordinaire.

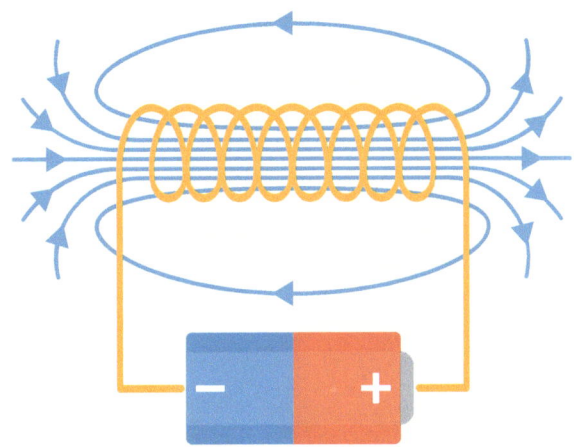

The electric motor he also designed.
The concept has since been endlessly refined.

He studied the magnetic effects on light.
The electromagnetic theory owes him for his insight.

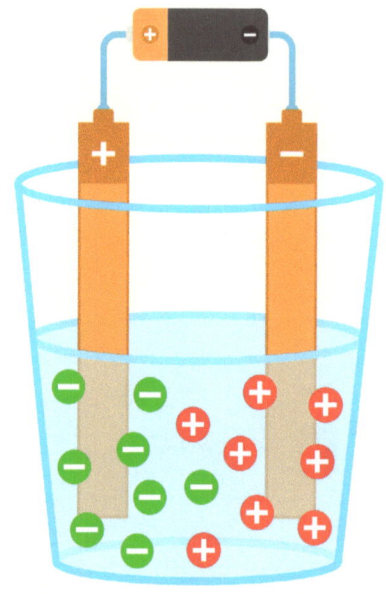

He caused an electric current to induce chemical change.
From physics to chemistry his accomplishments had quite a
range.

James Maxwell

In physics, the unification of laws is greatly prized.
Newton got there first, but with Maxwell
physicists were pleasantly surprised.

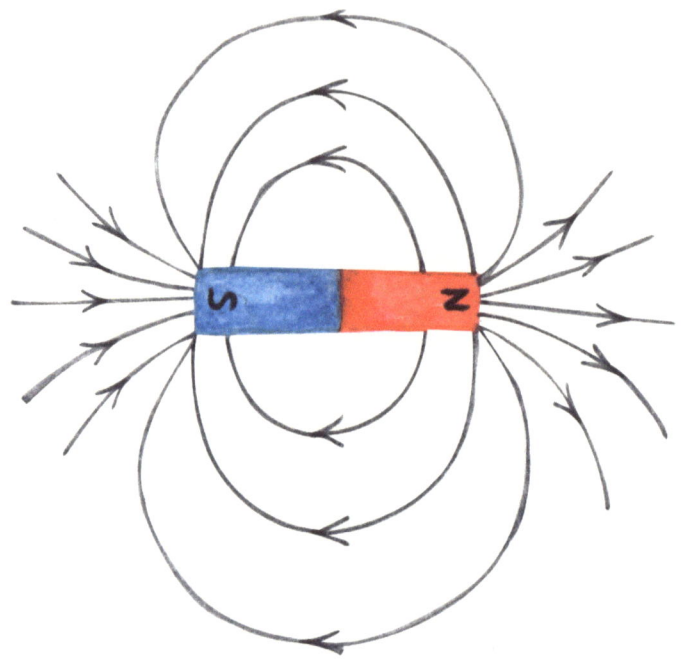

Electricity and magnetism were going their own way,
but James Maxwell had the final say.

Electricity and magnetism are fundamentally the same.
The difference is only in their name.

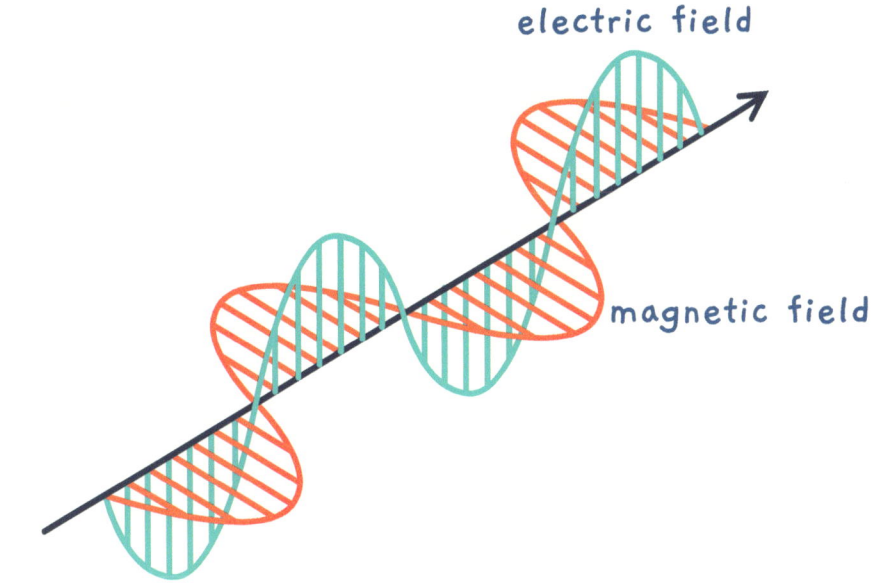

electric field

magnetic field

The electromagnetic nature of light
was revealed by Maxwell's brilliant insight.

$$\nabla \cdot E = \frac{\rho}{\varepsilon_0}$$

$$\nabla \cdot B = 0$$

$$\nabla \times E = -\frac{\partial B}{\partial t}$$

$$\nabla \times B = \mu_0 J + \mu_0 \varepsilon_0 \frac{\partial E}{\partial t}$$

He summed up his work into equations four,
a feat that had never been done before.

His name is taken along with Newton and Einstein.
Among great physicists, he stands at the front of the line.

17

Ludwig Boltzmann

The laws of thermodynamics dictate
if a body is in a hot or cold state.

Microstates in thermodynamics are the key.
They are so small that we can never see.

They are probabilistic and restless.
Entropy is the measure of that mess.

P(A | B)

Boltzmann thought this way.
Study probability, he would say.

His colleagues went on the offense.
He was distraught and put on the defense.

His life had a tragic fate.
He got recognized but it was too late.

Max Planck

At the end of the 1900's most of the physics
was thought to have been done.

There was a feeling that physics did
not offer any more excitement or fun.

They couldn't be more wrong.
Startling discoveries were to come along.

It started with how a heated body radiates.
There are limits to how much energy it generates.

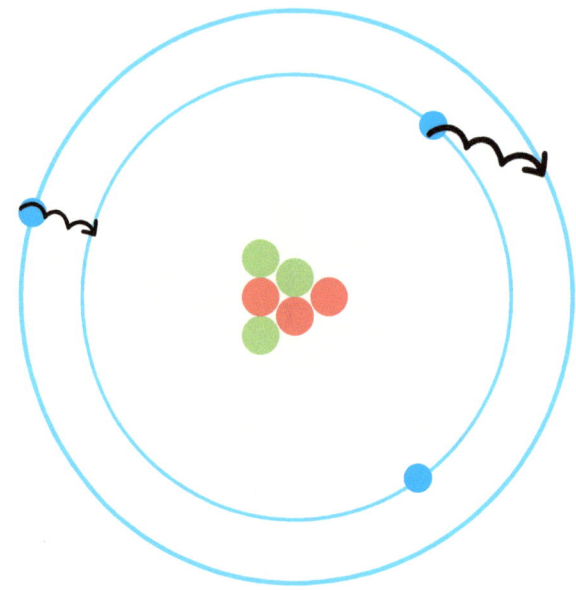

Planck postulated energy came out in chunks.
The energy levels were discrete with sudden jumps.

This led to Planck's recognition and fame.
The smallest units in physics still carry his name.

Marie Curie

In so many ways she was a pioneer.
A life lived without any fear.

Against all odds she survived.
In a man's world she thrived.

24

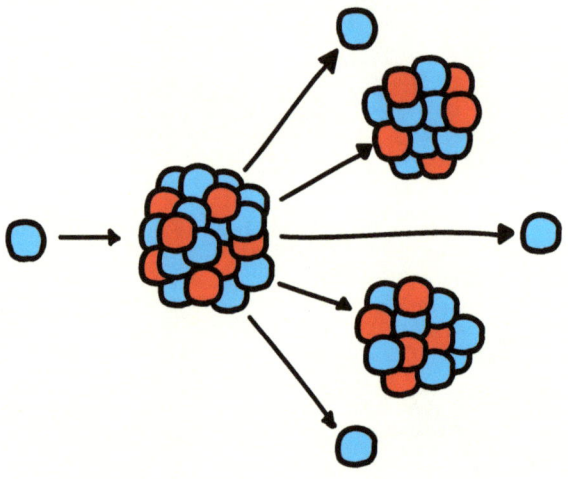

She worked along with her spouse
living in a truly remarkable house.

She discovered radium that glowed.
The seeds of radioactive physics were sowed.

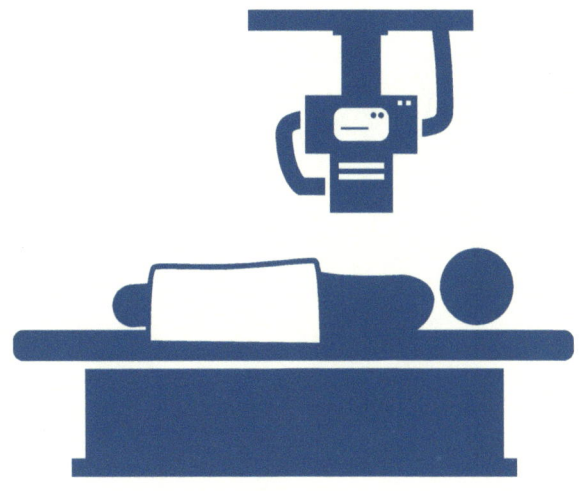

She knew how to use X-rays.
And helped the war injured in those days.

She got the Nobel Prize twice.
Working with radioactivity though had its price.

Albert Einstein

We all know Einstein was smart.
But he didn't get an auspicious start.

In a patent office, he was stuck.
For an academic position, he was out of luck.

$$E = mc^2$$

But 1905 was his miracle year.
The path to special relativity was clear.

As things move, time slows.
An observer at rest already knows.

The rest energy is mass times the speed of light squared.
For the nuclear age, the world was not at all prepared.

Einstein was suspicious of Newtonian laws.
He went on a quest to find gravity's real cause.

In a thought experiment, a person was in free fall.
Einstein realized gravity was not acting at all.

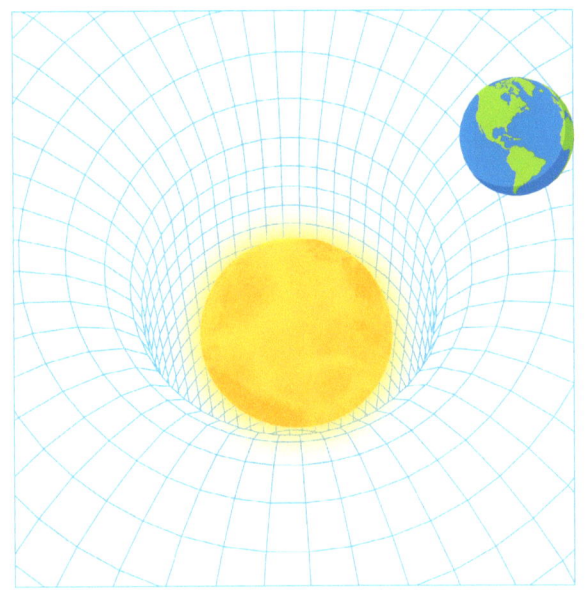

Acceleration and gravity were all the same.
The difference was only in their name.

Gravity was not a real force.
The curvature of spacetime was what laid the course.

30

This is when general relativity came into its stride.
The implications were far and wide.

When Hubble discovered the universe was drifting apart.
It meant the universe must have had a start.

Even in extreme cases, spacetime curvature plays a role.
And when a massive star collapses, we get a black hole.

Einstein showed us the power of human thought.
It's upon us if we do good or we cannot.

Ernest Rutherford

He conducted a legendary experiment.
Onto a gold foil, Helium nuclei were sent.

Most went through, only a few were bent.
Most of the atom was found empty, that's what it meant.

33

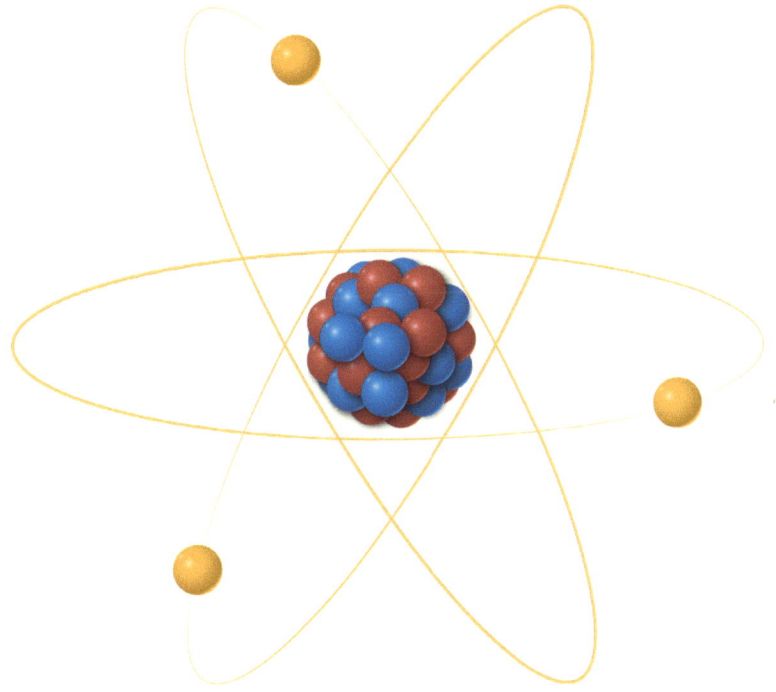

The nucleus of the atom forms a tiny core.
Where most of the mass is in store.

He explained half-life, alpha, and beta decay.
Why over time a radioactive atom gives it all away.

Niels Bohr

The structure of the atom to its core
was unveiled by none other
than the great Niels Bohr.

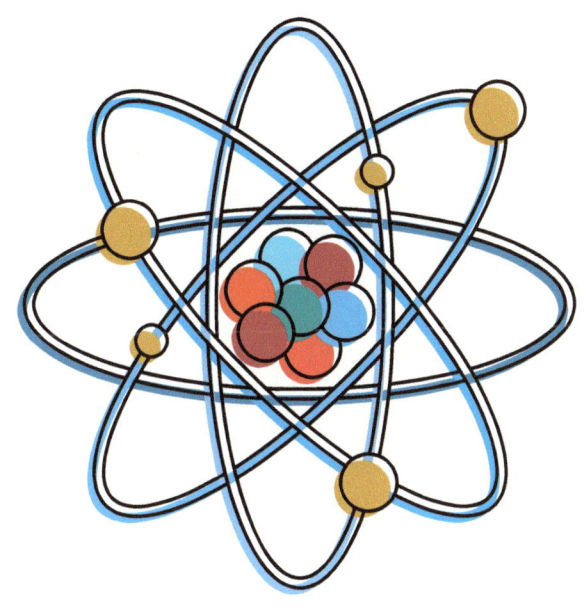

With the nucleus at the center,
many electrons were found,
and they moved round and round.

The atomic model was revolutionary and new,
Bohr became a towering personality and his
proteges were also quite a few.

Louis De Broglie

He came up with the idea of wave-particle duality.
All particles carry wave-like properties,
that's the reality.

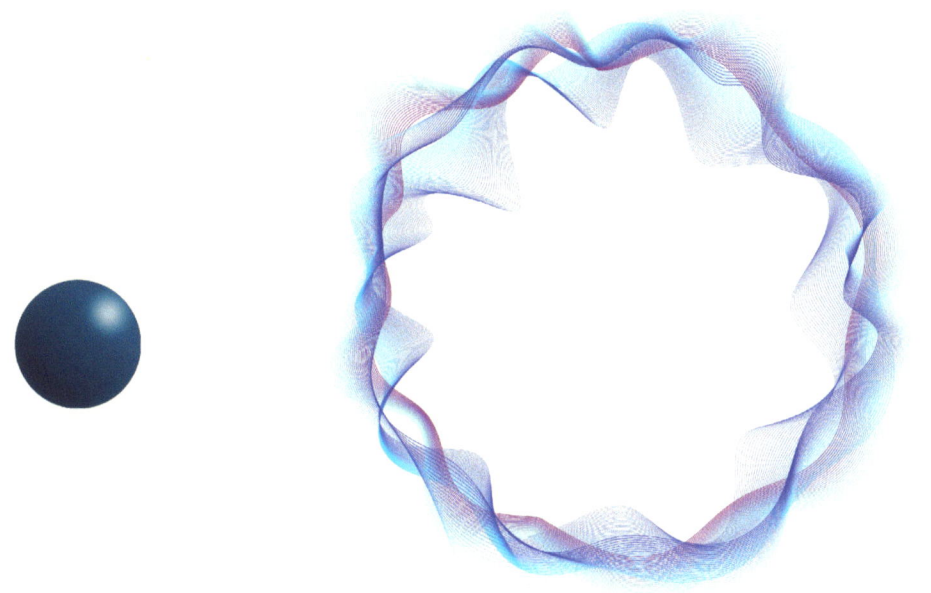

A particle is like a dot,
but a wave is definitely not.

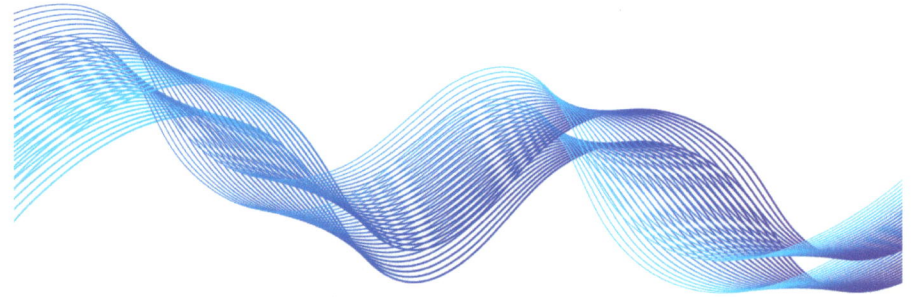

Particles get hit and scatter.
For waves, that doesn't even matter.

Waves can cancel and add.
Particles and waves do coexist,
that's the insight De Broglie had.

Wolfgang Pauli

Pauli sealed the electron's fate.
They cannot occupy the same quantum state.

As electrons move around,
many atomic orbitals are found.

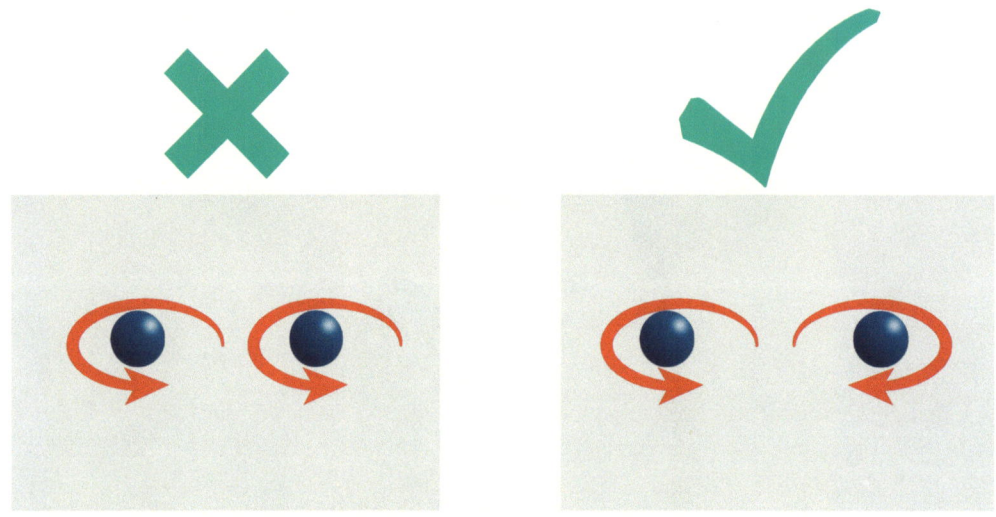

An orbital can have two electrons,
but with spins that cannot be aligned.
That's how the Pauli exclusion principle is defined.

Erwin SchrÖdinger

Equations help us predict,
to navigate the laws of physics,
that are more or less strict.

$$i\hbar \frac{\partial \Psi}{\partial t} = H\Psi$$

Things in quantum mechanics are the same.
The fundamental equation of quantum mechanics
bears SchrÖdinger's name.
Things happen at random in quantum mechanics,
that's quite true.
But the SchrÖdinger equation makes predictions,
that are also quite a few.

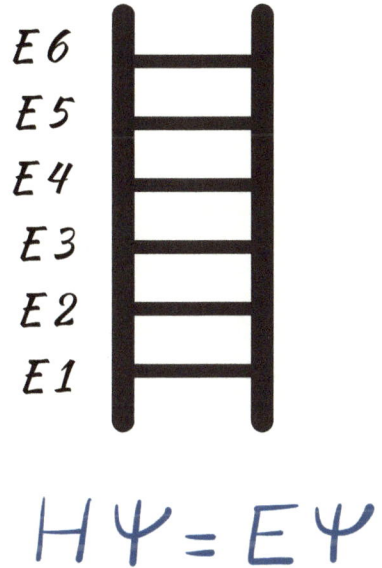

$$H\Psi = E\Psi$$

Where can electrons be found?
What are the energies that are bound?
The predictions of the Schrödinger equation
remain valid and sound.

If you put a cat in a box with a random gun, SchrÖdinger said.
Unless you open the box, the cat is both alive and dead.
He recognized quantum mechanics was weird.
How come the superposition disappeared?

Max Born

He took up a matter grave
to make sense of the quantum wave.
The wave function was an enigma.
Some even said it carried a certain stigma.

$$\Psi = $$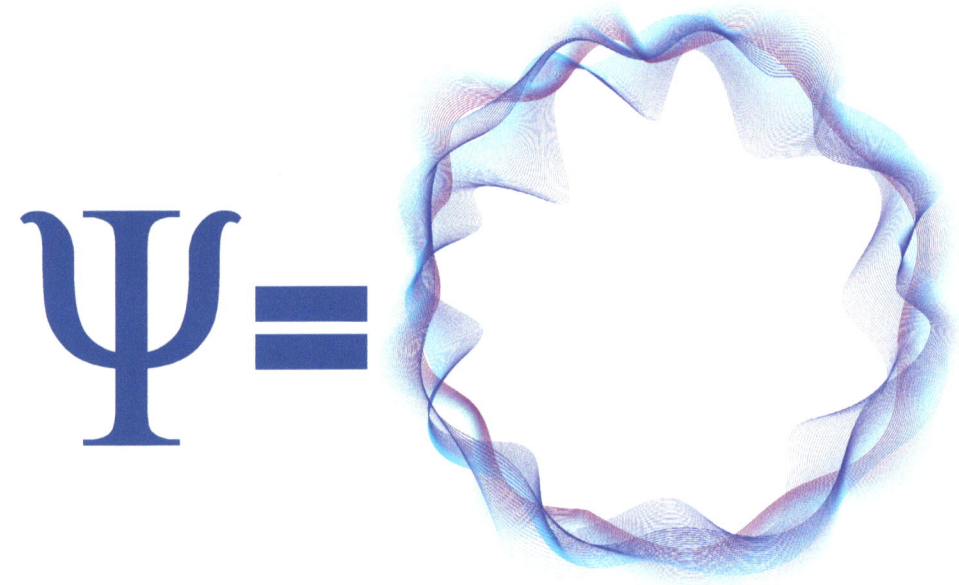

A particle is represented by this function.
It's a theory and reality meeting junction.

Max Born developed this curiosity
to know the meaning of this monstrosity.

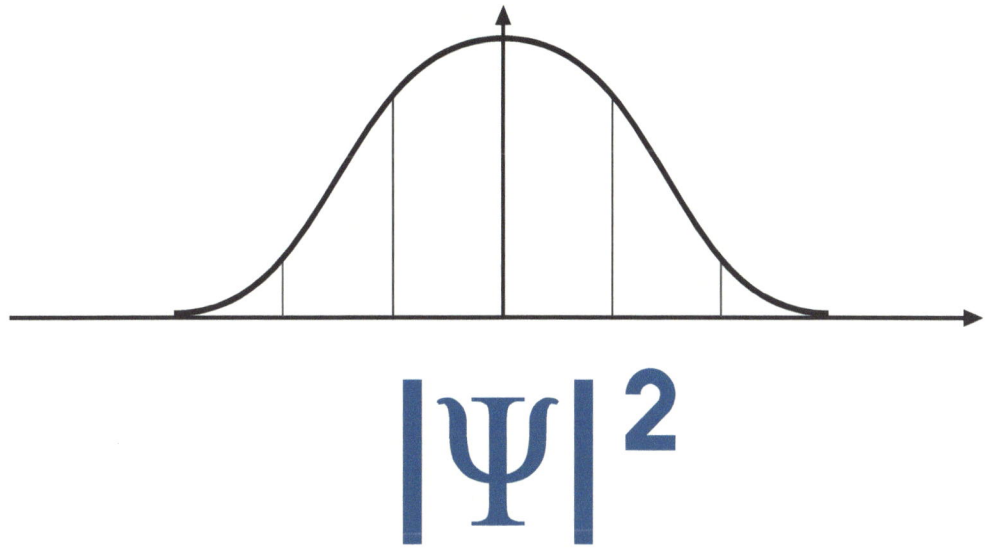

$$|\Psi|^2$$

He postulated that if the wave function is squared,
then the victory of knowing the whereabouts
of a quantum particle could be declared.

Werner Heisenberg

Want to know a particle's whereabouts?
You'll always be left with lingering doubts.

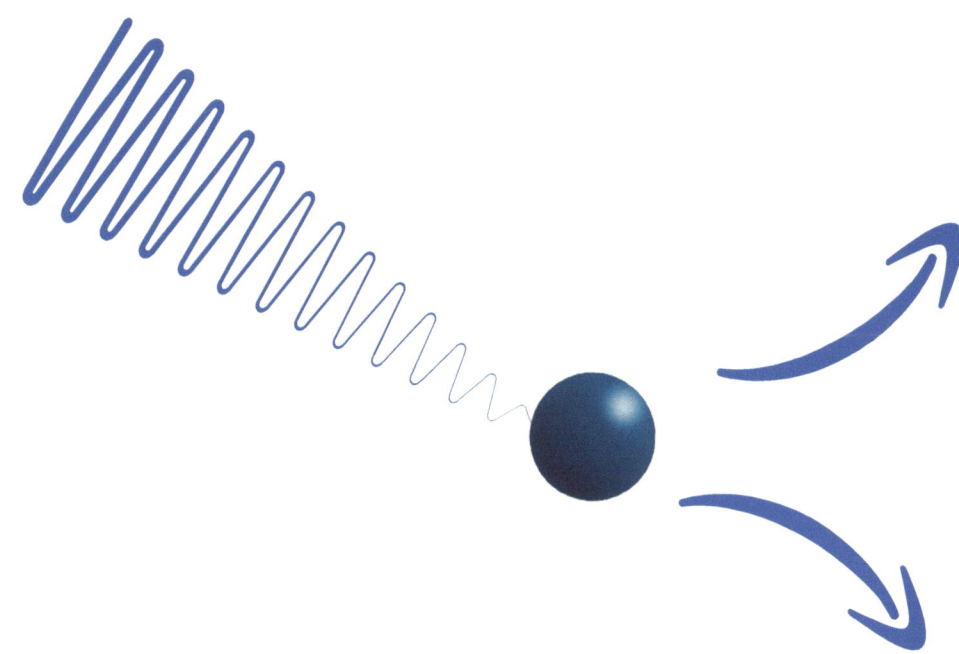

To know its position, you must shine a light.
But then you don't know where it's going
as it gets kicked in this fight.

If you know where it's going in this test,
then you have no idea about its position
even if you try your best.

If you get clever and fine-tune a bit,
you run into the uncertainty limit.

This is Heisenberg's uncertainty principle in all its glory.
And quantum reality will always be an incomplete story.

51

Paul Dirac

Paul Dirac was quiet and reserved.
His reputation as a gentle genius
is well deserved.

It was hard to carry a conversation
with him as such.
With very few words,
he still said so much.

Special Relativity   Quantum Mechanics

The speed of light was tough to catch.
Quantum Mechanics was no match.

How to get them close?
The path Paul Dirac chose.

matter

antimatter

It was not easy or fast,
but he came up with
the Dirac equation at last.

It predicted electrons with
spin up and down.
But the prediction of antimatter
was where it got the crown.

Subrahmanyan Chandrasekhar

Black holes suck.
Forget the stars and planets,
even the light gets stuck.

Black holes were thought to be rare.
As if the universe didn't really care.

55

When a star's fuel runs out,
it collapses under the gravity without a doubt.

The particle pressure offers some resistance from the inside.
But beyond a critical mass it can't stop the gravity's tide.

The Chandrasekhar limit is that critical state
where a neutron star or a black hole becomes a star's
inevitable fate.

Enrico Fermi

He was the main architect of the nuclear age.
In both theory and experiments, he continued to amaze.

The nuclear power he knew how to harness.
His grasp of statistical mechanics was no less.

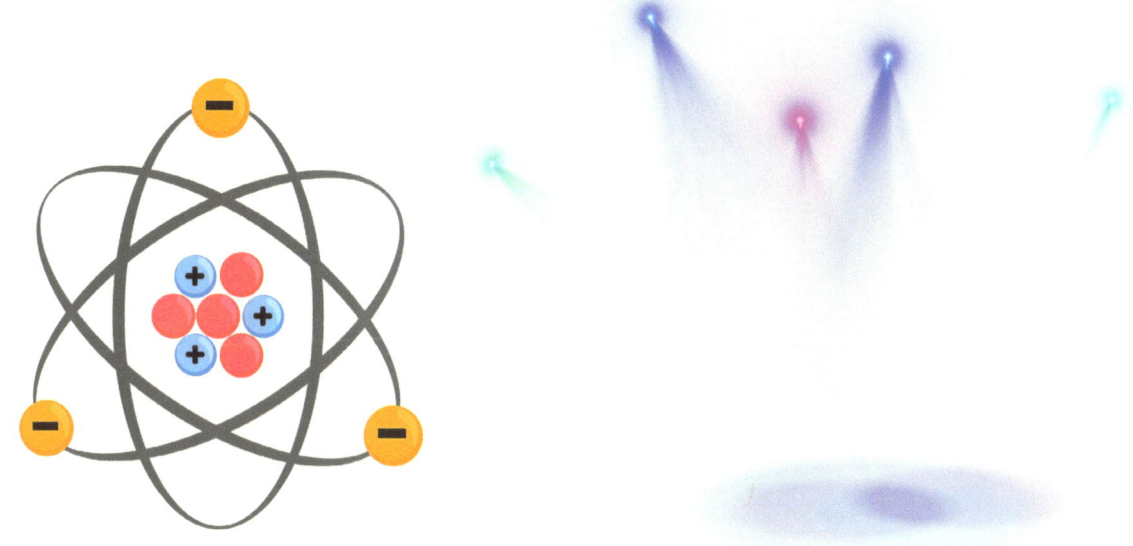

Fermions(electrons, protons, neutrons)   Bosons(photons of light)

All matter is made of fermions that carry his name.
The force carrying bosons don't behave the same.

Considering his grasp of the subject,
he was also chosen for the Manhattan Project.

After the war, his research continued in high gear.
In high-energy physics, he remained a pioneer.

Richard Feynman

He may have been flamboyant and loud.
But he knew how to stand out from the crowd.

A master in theory and communication.
He developed the quantum path integral formulation.

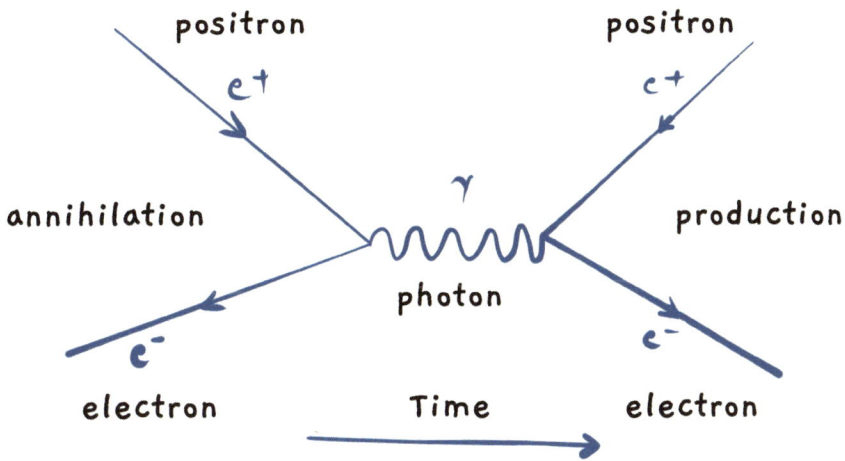

He developed the theory of electrons and light.
Superfluidity and particle physics were also in his sight.

He developed diagrams on how particles scatter.
To simplify solving this important matter.

Feynman lectures were a work of art.
It gave many budding physicists a head start.

Murray Gell-Mann

The structure of the proton was hard to crack.
Most physicists were on the wrong track.

The problem was that protons couldn't be cut.
Physicists were wondering if doors were forever shut.

Murray Gell-Mann came up with an ingenious way.
Physicists paid attention to what he had to say.

He said the proton is made of quarks with colors three.
They are always bound and cannot be made free.

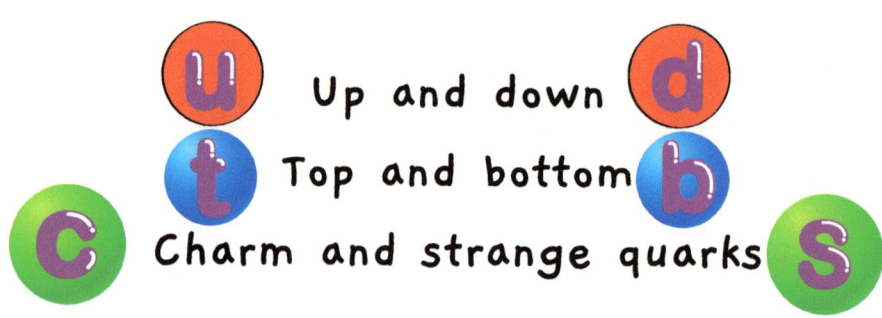

Up and down
Top and bottom
Charm and strange quarks

As if quarks follow the rules of the heart.
Their bonds get stronger if taken apart.

Quarks come in different flavors too.
And the total types of quarks are also quite a few.

Steven Weinberg

Unification of laws plays a central role.
If simplification of physics is the goal.

There are four forces in the universe to unite.
Gravity, strong and weak nuclear force and
the electromagnetic light.

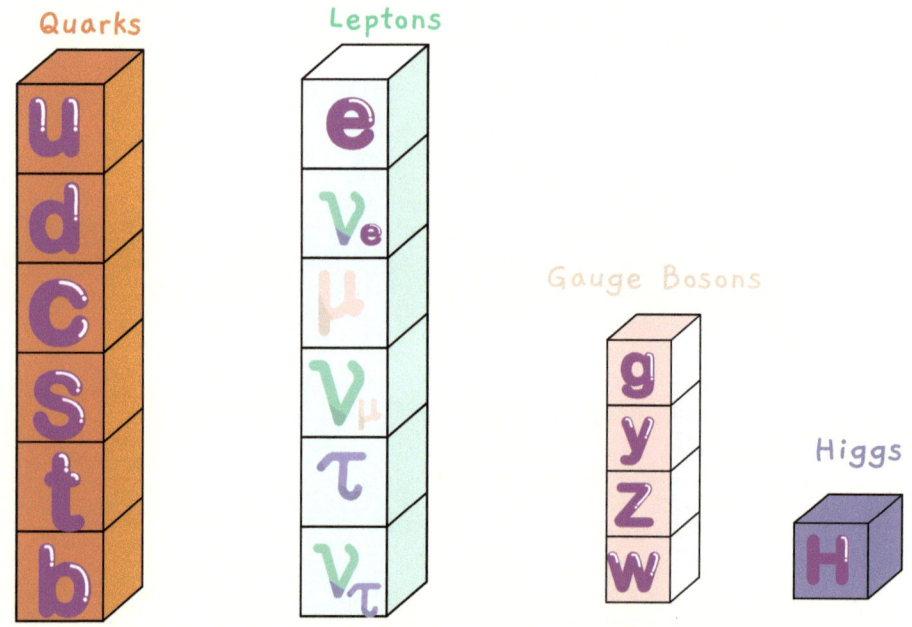

The standard model of particle physics
encompasses all that we know.
Unification of forces it must show.

Maxwell unified electricity and magnetism.
The search was on for some other mechanism.

Strong force

Gravity

Electroweak force

Weinberg, Salam, and Glashow were key
in developing the electroweak theory.

Gravity and strong force unification still remains elusive.
There are many proposals, but nothing is conclusive.

About the Author

Preetinder Rahil
writes fiction, non-
fiction, and poems
that rhyme. He tries
to keep things
simple, fun, and
worth your time.

www.ingramcontent.com/pod-product-compliance
Lightning Source LLC
Chambersburg PA
CBHW041501280526
45792CB00004B/1084